Fatiha Arioui
Djamel Ait Saada
Abderrahim Cheriguene

Orange pectin and yoghurt quality

Fatiha Arioui
Djamel Ait Saada
Abderrahim Cheriguene

Orange pectin and yoghurt quality

ScienciaScripts

This book is a translation from the original published under ISBN 978-3-8416-1397-4.

Publisher:
Sciencia Scripts
is a trademark of
Dodo Books Indian Ocean Ltd. and OmniScriptum S.R.L publishing group

120 High Road, East Finchley, London, N2 9ED, United Kingdom
Str. Armeneasca 28/1, office 1, Chisinau MD-2012, Republic of Moldova, Europe
Printed at: see last page
ISBN: 978-620-7-24531-4

Pectin and yoghurt quality

Author : FATIHA ARIOUI *

Co-authors: DJAMEL AIT SAADA and ABDERRAHIM CHERIGUEN E

Affiliation: Laboratoire de Technologie Alimentaire et
Nutrition, Abdelhamid Ibn Badis University, Mostaganem, Algeria.

*arioui-fatiha@hotmail.fr ; fatiha.arioui@univ-mosta.dz

Tel: +213 558888319.

TABLE OF CONTENTS:

Introduction

Pectin is a macromolecule mainly extracted from apple pomace and citrus peel. It is used in a wide range of applications because of its image as a natural product, its nutritional properties and, above all, its functional properties **(Mesbahi et al., 2005; Combo et al., 2011)**.

In the food industry, it is common practice to add stabilising agents to yoghurt mixes. Their purpose is to improve and maintain the desired characteristics of the final product, such as a certain firmness, viscosity or consistency, a suitable texture and a pleasant mouthfeel. The stabilisers generally used are pectin, gelatine and whey proteins **(Sodini et al., 2004)**. Depending on their functional properties and concentrations, they can gel or increase the viscosity of the gel. As a result, they help to limit syneresis and give a smooth texture.

Yoghurt is the most widely consumed fermented milk in the world, thanks to its highly sought-after organoleptic and nutritional properties **(Loveday et al., 2013; Sahan et al., 2008)**. It is obtained by lactic fermentation of just two specific strains: *Steptococcus thermophilus* and *Lactobacillus bulgaricus* **(Kumar and Mishra. 2004; Sokolinska et al., 2004)**. The health benefits of this product, its range of flavours and the arrival of probiotic-enriched yogurts are responsible for the increase in its consumption. Nevertheless, the main defects of this fermented product remain inadequate gel firmness, variations in viscosity and the expulsion of whey: syneresis **(Keogh and O'Kennedy, 1998; Lucey et al. , 1998)**.

The current dynamism of the foodstuffs market is forcing manufacturers to constantly formulate new products, especially fermented dairy products (yoghurt), consumption of which has grown considerably as a result of technological progress, which has made it possible to improve the specific aspects of these products through the use of ingredients such as texturing agents, thickeners and gelling agents **(Kar and Arslan, 1999)**.

Yoghurt is obtained by coagulating milk without draining. It is manufactured according to procedures that differ depending on the nature of the finished product (stirred, firm), but which always involve lactic fermentation leading to gelation due to the destabilisation of the protein system under the effect of a drop in pH. Texture is an essential component of yoghurt quality. It is influenced by a number of factors: the composition of the milk, the extent of the treatments applied before manufacture (heating, homogenisation) and the cultural conditions (ferment, temperature).

The manufacture of yoghurts can therefore result in certain technological problems such as inadequate gel firmness, variations in viscosity and whey exudation. To prevent these quality defects, yoghurts are usually enriched with texturising, stabilising and gelling agents. One of these food additives is pectin.

CHAPTER 1

1. Pectins

1.1. Definition

Pectins are complex polyosides used in the cell walls of most higher plants. They are mainly present in the middle lamella and the primary wall **(Jenson et al., 2008)**. The quantity of pectic substances in plants varies greatly depending on their botanical origin and history (cultivation method, growth period, etc.) **(Caffall and Mohnen, 2009)**.

Pectins are abundant in fruit and vegetables and change as the tissues ripen. Although they can be extracted from a large number of plants, the main industrial sources of pectin are apple pomace and lemon and orange peel **(Pinheiro et al., 2008; Combo et al., 2011)**.

Pectin production is the most sensible way of using by-products from the citrus juice industry, both economically and ecologically. Apple pomace contains between 10% and 15% pectin and orange peel contains around 20% to 30% **(Srivastava and Malviya, 2011)**.

Pectins are essentially made up of galacturonic acid (Gal A) residues linked together by a (1→4) linkages and partially acetylated or esterified by methyl groups **(Ridley et al., 2001)**.

Pectin is used in the food industry as a gelling additive, thickener, stabiliser and emulsifier. It is also used in the composition of pharmaceutical specialities for its anti-acid, haemostatic or anti-diarrhoeal properties **(Wicker et al., 2014)**.

1.2. Pectin production processes

Pectin is mainly extracted industrially from by-products of the fruit juice industry. The pectin extraction process is based on hydrolysis of the raw material in an acid medium; this stage is generally carried out at 90°C for a minimum of one hour **(Iglesias and Lozano, 2004)**, followed by filtration and precipitation in alcohol **(Kalapathy and Proctor, 2001)**.

These conditions are particularly detrimental to the quality of the pectin. The coagulum, which has a fibrous appearance, is washed, pressed, dried under vacuum and then ground to obtain a fine powder.

Some research has focused on pre-treating the raw material prior to hydrolysis to facilitate the hydrolysis of protopectin (which is strongly bound to cell walls) and to increase pectin extraction yields. **Wang et al (2007)** showed that using microwaves to extract pectins would reduce extraction times and therefore costs. **Fishman et al (2000) reported that** microwave pre-treatment of various fruits considerably increased pectin yields.

1.3. Structure and characteristics of pectin

These substances are exclusively of plant origin **(Ridley et al., 2001)**. They are polysaccharides made up of a main galacturonic acid chain and a branched secondary chain **(Figure 1)**. The main chain is made up of D-galacturonic acid (D-Gal A) linked by a (1→4) glycosidic bonds. Regularly between these monomers rhamnose molecules are installed by a (1→2) linkages **(Wicker et al., 2014)**.

Pectins have specific physico-chemical properties due to their poly-electrolyte nature. These characteristics give them the ability to associate with each other and form gels in the presence of divalent cations such as calcium, making them susceptible to interactions with proteins.

Pectins are characterised by their degree of methylation (DM). The degree of methylation is a very important parameter that influences the process and mechanism of association of pectins in the formation of gels. The degree of methylation is the percentage of carboxyl groups esterified by methanol **(Lira-Ortiz et al., 2014)**.

According to **Willats et al (2006), there** are two types of pectin depending on the degree of methylation (DM):

- HM (highly methylated) pectins: these are pectins in which the degree of esterification is greater than 50%.
- LM pectins (weakly methylated): these are pectins whose degree of esterification is less than 50%.

Pectin is rich in D-galacturonic acid and is classified into four main groups: Homogalacturonan (HG), Rhamnogalacturonan I (RGI), Rhamnogalacturonan II (RGII) and Xylogalacturonan (XGA).

- **Homogalactiironane (HG): consists of** a linear chain of residues a-1,4-Gal A, which is often methylesterified at the C-6 position and can be acetylated at the O-2 and/or O-3 position **(Jensen et al., 2008; Atnodjo et al., 2013).**
- **Rhamnogalacturonan I (RGI): consists of** a backbone with the disaccharide [a-1, 4-Gal A- a-1, 2- Rha] as the basic repeating unit. The rhamnose residues in RGI are often substituted with galactan, arabinose, or arabinogalactan I side chains **(Jensen et al., 2008; Wicker et al., 2014).**
- **Rhamnogalacturonan II (RGII)**: is a polysaccharide with a complex structure that appears to be remarkably conserved throughout vascular plants. RGII consists of a short chain of Homogalacturonan substituted with four different side chains and is composed of 12 different monosaccharides in more than 20 different linkages **(Mohnen, 2008).**
- **Xylogalacturonan (XGA): consists of** a Homogalacturonan (HG) backbone substituted with single residues of P-1,3- Xly or such residues are substituted with a few additional xylose residues **(Jensen et al., 2008).**

5

1.4. Properties of pectins

Pectin is a non-digestible polysaccharide classified as a dietary fibre. This polysaccharide has many uses in the food industry (as a thickener, stabiliser or gelling agent). In addition, pectin has a very high water retention capacity. These functional properties depend on the structure of the pectin molecule and, above all, its degree of methylation.

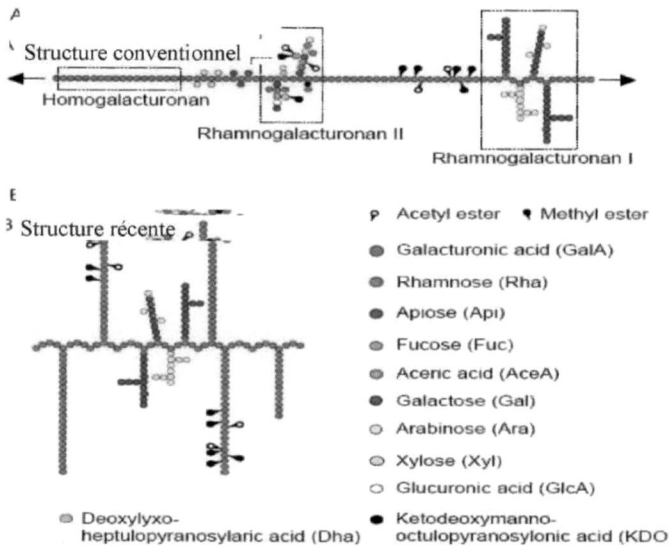

Figure 01. Basic structure of pectin (Willats et *al.*, 2006).

The degree of methylation is an important parameter influencing the process and mechanism of pectin association in gel formation (**Combo et al., 2011**). Highly methylated (HM) pectins form gels in the presence of sugar and in acidic media. A long-standing application for this type of pectin is in jam formation. Low-methyl pectins (LM) form gels in the presence of bivalent ions, such as calcium ions. Apart from the degree of methylation, the sugar or acid concentration, the presence of side chains and the temperature also play an important role in the consistency of the gel (**Srivastava and Malviya, 2011**).

These properties give pectin a considerable importance that is much sought after in the various fields of use: the food, pharmaceutical, biotechnology and pollutant treatment industries (**Wicker et al., 2014**).

1.5. Main uses of pectin

The general tolerance of pectin as a natural substance and the fact that it is a soluble dietary fibre make it suitable for use in a variety of applications.

Pectin is widely used in the food industry as a thickener and emulsifier in dairy products and as a gelling agent and colloid stabiliser in jams and jellies.

From a nutritional point of view, pectins are considered to be soluble dietary fibres with a high water retention capacity. They are used as dietary fibres and have physiological effects on the intestinal tract by reducing transit time and glucose absorption **(Olano-Martin et *al.*, 2002)**.

The use of pectins in the development of oligosaccharides for prebiotic and pharmaceutical applications is an emerging field. There is great interest in pectic oligosaccharides (POS) because of their potential for use as a new generation of prebiotics. Some of the qualities attributed to these oligosaccharides are protection against colon cancer and antibacterial activity. Repression of lipid accumulation in the liver, inhibition of bacterial adhesion to epithelial cells and stimulation of the growth of bifidobacterium and eubacterium rectale **(Olano-Martin et *al.*, 2002; Combo et *al.*, 2011)**.

A number of health benefits have been reported in terms of reduced retention of heavy metals in the body **(Combo et *al.*, 2011)**

CHAPTER 2

2. Methodology

This work is the fruit of two years of laboratory work, which began in April 2014 and was completed in May 2016. The experiments were carried out at the Food Technology and Nutrition Laboratory (TAN) of the Faculty of Natural and Life Sciences at the Abdelhamid Ibn Badis University in Mostaganem and the Food Microbiology and Biochemistry Laboratory of the Faculty of Sciences at the Hassiba Ben Bouali University in Chlef.

2.1. Raw materials

The raw material used to extract the pectin was orange peel of the *Citrus sinensis* L variety collected in the Chlef region in December 2014. The peels were separated from the endocarp, which represents 28% by mass of the fruit. The skins were dried at 50°C in an oven and then placed in airtight bags until further use.

The skimmed milk (spry skimmed milk powder, Belgomilk CVBA-Belgium) used in our experiment was supplied to us by a dairy located in Mostaganem -Algeria-.

2.2. Lactic ferments

The two yoghurt-specific lactic ferments used in the experimental study, *Streptococcus thermophilus* (YC-X16) and *Lactobacillus bulgaricus* (CHN-11), come from CHR (HANSEN Denmark) and are packaged in freeze-dried form.

2.3. Pectin extraction process

Pectin was extracted using the method of **Rezzoug et al (2008)**. Pectin was extracted from *Citrus sinensis* orange peel in a hot acid solution, then precipitated in a 96% alcohol solution° **(Figure 02)**.

The dried orange peel was crushed for 20 seconds, and the crushed peel (10g) was added to a 200ml solution of 0.1N hydrochloric acid (HCl), boiled in a reflux system at 90°C for 40 min and then placed on ice to stop the hydrolysis process.

The supernatant was recovered after filtration. The pectin was then precipitated with two volumes of 96° alcohol to one volume of supernatant. The precipitate obtained is washed with one volume of 96° alcohol. The pellet was collected, dried and finally ground to a powder. The pectin yield is expressed in g/100g of dried orange peel.

The pectin yield is calculated according to the following equation:

$$\mathcal{R}_{pectine} (\%) = 100 \text{ x } P / E$$

Where (R_{pectin}) is the percentage yield of extracted pectin, (P): the weight of extracted pectin,

8

and (E): the weight of dried orange peel used during extraction.

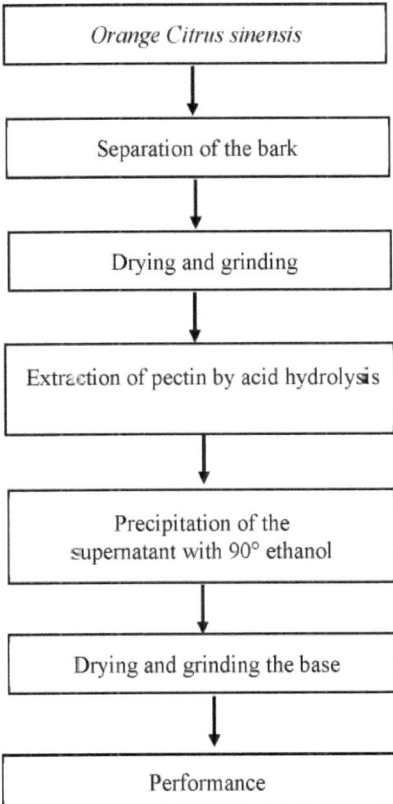

Figure 02. Pectin production diagram (Rezzoug et al., 2008).

2.4. Preparing sourdough

The milk used to make the sourdough is reconstituted from skimmed milk powder, prepared at a rate of 130 g/l.

The mixture is homogenised until it is completely dissolved, then thermised at a temperature of 100°C for 2 minutes in order to destroy the pathogenic germs and reduce the number of banal germs in the milk. The milk was then cooled to 45°C and inoculated with freeze-dried pure lactic strains specific to yoghurt at a rate of 0.25g/l of *Lactobacillus bulgaricus* (CHN-11) and 0.5g/l of *Streptococcus thermophilus* (YC-X16).

After activation of the strains by simple steaming at 45°C, the lactic sourdough was recovered and oriented for use in the various experimental milk production processes. The sourdough prepared had a strain ratio of two times *Streptococcus thermophilus* to one *Lactobacillus bulgaricus* (**Figure**

9

03).

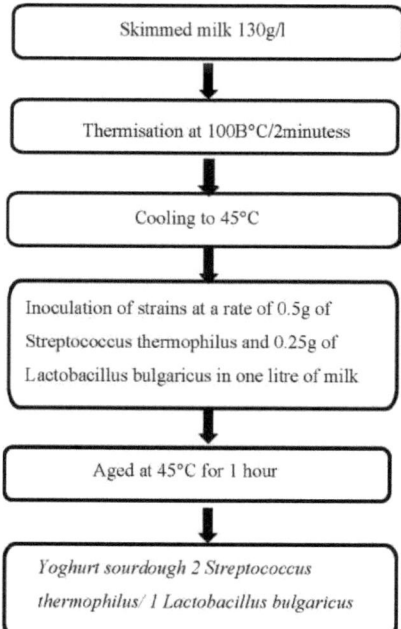

Figure 03. Stages in the preparation of lactic sourdough.

2.5. Process for making experimental yoghurts

The reconstituted milk used was prepared with 140 g/l of a 26% fat milk powder. The milk was then heat treated at a laboratory temperature of 100°C for 2 min to pasteurise it. Once cooled to 45°C, pectin was incorporated into the milk samples at different rates: 0, 0.1, 0.3 and 0.6%. Inoculation with yoghurt-specific lactic strains (CHR, HANSEN Denmark) was carried out in the trials at a sourdough rate of 3% and a ratio of *Streptococcus thermophilus* strains to *Lactobacillus bulgaricus* strains of 2S/1L.

Each experimental parameter was represented in triplicate in 3 x 100 ml jars. After steaming the samples at 45°C for 4 hours during the fermentation phase, the experimental fermented milks were cooled and stored at 4°C for 21 days during the post-acidification period.

2.6. Characterisation of pectin

2.6.1. Moisture and ash

The ash and moisture content of the pectin is determined using the **AOAC** method **(1980)**. Moisture is determined by drying 1g of pectin at 100°C for 4 hours.

The ash content is determined by incinerating 1g of the pectin in a muffle furnace set at 600°C

10

for 4 hours. The ash content is calculated using the following formula:

$$Ash = \frac{Mass\ of\ ash}{Pectin\ mass} \times 100$$

2.6.2. Determining the equivalent weight

The equivalent weight, methoxyl content and galacturonic acid content are determined using the method proposed by **Owens et al (1952)**. The equivalent weight value is used to calculate the galacturonic acid content.

The equivalent weight is determined by measuring 0.5g of pectin in a vial with 5 ml of ethanol. Next, 1g of sodium chloride (NaCl) is added to the solution. Next, 100ml of distilled water and 6 drops of phenol red (as a colour indicator) were added. The mixture was then shaken rapidly to ensure that the pectin was well solubilised.

The solution is titrated with NaOH (0.1N) until the colour of the indicator (phenol red) turns a persistent pink for at least 30 seconds. The neutralised solution is used to determine the methoxyl (MeO) content. The following equation is used to calculate the equivalent weight:

$$Equivalent\ weight = \frac{Sample\ mass \times 1000}{Volume\ of\ NaOH \times NaOH\ normality}$$

2.6.3. Methoxyl content

The methoxyl (MeO) content was determined by adding 25ml of 0.25N NaOH to the neutralised solution with stirring, then leaving the solution to stand for 30 min at room temperature. Next, 25ml HCl (0.25N) is added and the solution is titrated with NaOH (0.1N) until the colour of the indicator (phenol red) turns pink. The following equation is used to calculate my methoxyl content:

$$MeO\ (\%) = \frac{meq\ of\ NaOH \times 31 \times 100}{sample\ mass\ (mg)}$$

Where: meq: equivalent meli ;

    31: molecular weight of methoxyl (MeO).

2.6.4. Galacturonic acid content

The galacturonic acid content is calculated using the equivalent weight value and the methoxyl (MeO) content according to the following equation:

$$AG\ (\%) = \frac{176\ (meq\ NaOH\ for\ free\ acid + meq\ NaOH\ for\ saponification) \times 100}{sample\ mass\ (mg)}$$

11

Where: **176**: molecular weight of uronic acid.

2.6.5. Degree of esterification

The degree of esterification of pectin (DE) is calculated as follows:

$$\mathbf{DE}\ (\%) = \frac{176 \times \mathbf{MeO}\ (\%) \times 100}{31 \ \times \mathbf{AU}\ (\%)}$$

Where: **MeO**: methoxyl content

      AU: uronic acid content.

2.7. Physicochemical and microbiological analyses of yoghurts with added pectin

Physico-chemical and microbiological analyses were carried out at 2 h and 4 h during the fermentation period and weekly during the post-acidification period, while the samples were kept cold at 4°C for 21 days.

2.7.1. Physicochemical analyses

Physico-chemical analyses are carried out using the **AOAC (2005)** method.

a. pH and acidity

The pH is measured using a pH meter calibrated with two solutions: one basic and the other acidic at a temperature of 25°C.

Dornic acidity is determined by titrating 10 ml of an experimental fermented milk sample with 0.1 N NaOH in the presence of phenolphthalein as a colour indicator. The results are expressed in degrees Dornic **(AFNOR, 1980)**.

b. Viscosity

Viscosity is determined using a falling ball viscometer. The viscosity is determined using a glass tube 18 mm in diameter and 18 cm long fitted with a stopwatch and a standardised ball. The viscosity is expressed in Pascal seconds (Pas) according to the following equation:

$$\eta = \frac{(\rho'-\rho).g.D^2}{18.\nu} \quad \text{or even} \quad \eta = K.(\rho'-\rho).t$$

η: dynamic viscosity (Pas) ;

ρ': density of the ball (gm-3) ;

ρ: density of experimental yoghurt (gm-3);

g: the force of gravity (9.81 ms);$^{-2}$

D: diameter of the ball (m) ;

υ: ball speed (ms-1).

2.7.2. Microbiological analyses

Streptococcus thermophilus and *Lactobacillus bulgaricus* were counted using the method described by the **International Dairy Federation (IDF Standard 306) (2003)**. The M17 medium is used for the enumeration of *Streptococcus thermophilus* and the MRS medium for the enumeration of *Lactobacillus bulgaricus*. Results are expressed as colony-forming units (CFU) per millilitre of product sample.

2.8. Organoleptic test

Every 7 days during the post-acidification storage period at 4°C, the organoleptic quality of the experimental fermented milks was assessed by a tasting panel using a rating scale of 1 to 10. The organoleptic test consists of assessing the experimental products according to several parameters: taste, cohesiveness, stickiness, aftertaste, odour and whey exudation.

Taste: This consists of assessing the extent of the acidity developed by the lactic germs seeded in yoghurt-type fermented milks during storage.

Cohesiveness: Determines the sample's maximum capacity to deform before breaking when crushed between the fingers.

Adhesiveness: Expresses the intensity of the interfacial forces developed between the surface of the coagulum and the surface of a spoon when the product is taken.

Aftertaste: The panellist is asked to assess the possibility of an aftertaste.

Whey exudation: This involves assessing the quantity of whey exuded.

Odour: the panelist is asked to detect whether or not there is a sensation of bad odour emanating from the product tasted.

2.9. Statistical processing

The results of the physico-chemical and microbiological analyses were statistically processed by a bi-factorial analysis of variance in total randomisation, followed by a comparison of the means two by two according to the NEWMAN and KEULS test. On the other hand, those relating to the organoleptic test were processed using Friedman's non-parametric test **(Stat Box 6.4)**.

CHAPTER 3

3. Results

3.1. Characteristics of pectin

The extraction yield of pectin in our study was estimated at 24.33% ± 0.5

The moisture content of commercial pectin (12.67% ± 1.60) is higher than that of orange peel pectin (11.52% ± 0.22). The ash content of orange peel pectin is evaluated at (9.00% ± 1.00%), whereas that of commercial pectin is of the order of (11.33% ± 0.57%).

The equivalent weight of commercial pectin is higher ($p<0.05$) than the equivalent weight of pectin obtained after orange peel extraction; 8492.09 and 620.03 respectively. The methoxyl content of orange peel pectin (1.73%) was significantly lower ($p<0.05$) than that of commercial pectin (2.02%). As for the degree of esterification, the values for *Citrus sinensis* pectin are lower ($p<0.01$) than commercial pectin at 28.79% and 82.03% **(Table 01)**.

Table 01: Characteristics of *Citrus sinensis* pectin and commercial pectin

	Orange peel pectin	Commercial pectin
Ash content (%)	9.00 b±1.00	11.33a ± 0.57
Humidity (%)	11.52b ± 0.22	12.67a ± 1.60
Equivalent weight	620.03b ± 21.75	8492.06a ±1435.17
MeO (%)	1.73b ± 0.22	2.02a ± 0.13
DE (%)	28.79b ± 1.89	82.03a ± 3.74
AG(%)	39.89a ± 0.73	11.96b ± 1.05
Colour	White yellow	White

The results are expressed as the mean followed by the standard error; a, b: statistical comparison of the means in pairs; GA: galacturonic acid; DE: degree of esterification; MeO: methoxyl.

Galacturonic acid, which indicates the degree of purity of the pectin, shows that the pectin extracted from *Citrus sinensis* bark, which was the subject of the study, contained 39.86% more galacturonic acid ($p<0.01$) than commercial pectin (11.96%). These results show that the purity of pectin extracted from *Citrus sinensis* bark is higher than that of commercial pectin **(Table 01)**.

3.2. Physicochemical quality of fermented milk with added pectin 3.2.1 pH

During the fermentation period, a clear decrease in pH values was recorded, with average values of 4.94 at 2 hours, and 4.57 after 4 hours of steaming. On the other hand, the decrease in pH during the post-acidification period was slow and gradual, with mean values varying from 4.13 to 4.12 and 4.05 at 7ème, 14ème and 21ème days, respectively **(Table 02)**.

14

Table 02: Evolution of the average pH content of yoghurts with added pectin

Periods		Pectin incorporated (%)				Effect of incorporating pectin
		0%	0.1 %	0.3 %	0.6 %	
Fermentation	2 H	5.34ᵃ ± 0.01	5.18 b ± 0.07	4.92ᶜ ± 0.02	4.92ᶜ ± 0.03	* *
	4 H	4.63ᵃ ± 0.01	4.65ᵃ ± 0.03	4.5 b ± 0.01	4.49 b ± 0.02	* *
	7 J	4.19ᵃ ± 0.03	4.13b ± 0.01	4.09 b ± 0.01	4.09 b ± 0.01	* *
Post-acidification	14 J	4.13 ± 0 03	4.15 ± 0.01	4.1 ± 0.01	4.11 ± 0.03	NS
	21 J	4.11 ± 0 03	4.04 ± 0.06	4.02 ± 0.01	4.02 ± 0.01	NS

The results are expressed as the mean followed by the standard error; ** Highly significant effect (p<0.01) of adding pectin; NS: non-significant effect (P>0.05) of adding pectin; a, b, c: statistical comparison of means two by two.

Throughout the fermentation period, and during the first week of postacidification, an inversely proportional relationship was established between the pH values of the experimental milks and the doses of pectin added (p<0.01).

For the second and third weeks of storage, the pH values develop independently of the pectin levels, stabilising at the final values (4.11, 4.04, 4.02, and 4.02) for pectin doses of (0, 0.1, 0.3 and 0.6%) incorporated into the products, respectively **(Table 02)**.

3.1.1. Acidity

During the fermentation period, the Dornic acidity of pectin-supplemented yoghurts showed a clear increase from an average of 70.25°D at 2 hours to 90.50°D after 4 hours of fermentation.

Table 03: Changes in Dornic acidity of yoghurts with added pectin

Periods		Pectin incorporated (%)				Effect of incorporating pectin
		0%	0.1 %	0.3 %	0.6 %	
Fermentation	2 H	68.33b ± 11.24	54.33ᶜ ± 3.21	70.5b ± 6.26	88ᵃ ± 1.73	* *
	4 H	80.66ᶜ ± 1.52	90.66b ± 1.52	91.66 b ± 5.68	99ᵃ ± 3.60	* *
Post-acidification	7 J	87.33 ± 5.77	92.66 ± 11.01	100.83 ± 3.32	99.66 ± 3.51	NS

	14 J	89.33 ± 6.65	96.66 ± 3.51	99.26 ± 4.19	99.66 ± 8.62	NS
	21 J	89.66 ± 7.09	98 ± 8.71	99.1 ± 7.53	100 ± 4.58	NS

The results are expressed as the mean followed by the standard error; ** Highly significant effect (p<0.01) of adding pectin; NS: non-significant effect (P>0.05) of adding pectin; a, b, c: statistical comparison of means two by two.

During the post-acidification period, a progressive increase in the acidity of the experimental fermented milks was recorded, ranging from 95.12°D at the start of storage, to an average of 96.23°D on day 14ème , reaching 96.69°D after 21ème days of cold storage of the samples at 4°C **(Table 03)**.

Furthermore, during the fermentation period, it appears that Dornic acidity is proportional to the increase in the pectin addition rate from (0, to 0.1, to 0.3 and to 0.6%) in the experimental yogurts (p<0.01); i.e. levels that vary from (68.33, to 54.33, to 70.5, and to 88°D) after 2 hours, and from (80.67, to 90.67, to 91.67, and to 99°D) after 4 hours of steaming, respectively.

The second post-acidification phase was marked by a slight increase in Dornic acidity over the three weeks of storage, with values varying from (87.33, to 92.67, to 100.83, and to 99.67°D) on the 7ème day, from (89.33, to 96.67, to 99.27, and to 99.67°D) on the 14ème day, and from (89.67, to 98, to 99.1, and to 100°D) on the 21ème experimental day, respectively for products with pectin added at (0, 0.1, 0.3 and 0.6%) **(Table 03)**.

3.1.2. Viscosity

Overall, throughout the fermentation period, the viscosity of the fermented milks tended to increase from (8.72 Pas) at 2 hours of fermentation of the products in the oven, to (23.82 Pas) on average after 4 hours, at the end of fermentation.

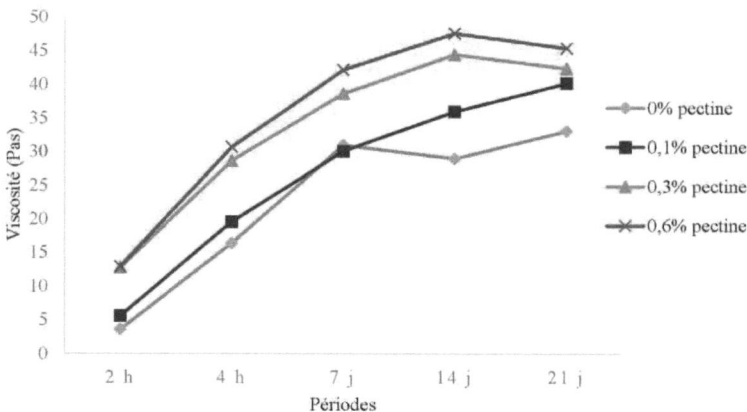

Figure 04. Changes in the average viscosity (Pas) of experimental fermented milks during the fermentation and post-acidification periods.

During the post-acidification period, viscosity increased in the same way, with mean values

16

varying from 35.47 to 39.23 and 40.23 Fas during the 7th, 14th and 21st days of storage of samples of cold fermented milk at 4°C **(Figure 04)**.

During the fermentation phase, the results show an increase in viscosity, positively correlated with the increase of (0, 0.1, 0.3, and 0.6%) in the levels of pectin incorporated into the fermented milks (p<0.01); i.e. respective values varying from 3.59, to 5.58, to 12.76, and to 12.93 Pas at 2 hours, and from 16.36, to 19.55, to 28.66, and to 30.68 Pas after 4 hours of steaming.

Table 04. Changes in the viscosity of yoghurts with added pectin

Periods		Pectin incorporated (%)				Effect of incorporating pectin
		0%	0.1 %	0.3 %	0.6 %	
Fermentation	**2 h**	3.59b ± 0.58	5.58b ± 2.11	12.76^a ± 1.37	12.93^a ± 1.96	* *
	4 h	16.36b ± 1.39	19.55^a b ± 5.29	28.66^a ± 7.26	30.68^a ± 5 24	*
Post-acidification	**7 J**	30.98b ± 0.67	30.11b ± 5.13	38.64^a ± 1.95	42.17^a ± 1.48	* *
	14 J	28.95^c ± 4.52	35.98b ± 3.79	44.42^a ± 2.34	47.56^a ± 1.84	* *
	21 J	33.05b ± 1.81	40.18^a ± 1.55	42.35^a ± 2.68	45.35^a ± 4.01	* *

The results are expressed as the mean followed by the standard error; ** Highly significant effect (p<0.01) of adding pectin; NS: non-significant effect (P>0.05) of adding pectin; a, b, c: statistical comparison of means two by two.

On the other hand, the effect of the addition of orange pectin on the viscosity of the products was highly significant (p<0.01) during the post-acidification cold storage period, with values varying from 28.95, to 35.98, to 44.42, and to 47.56 Pas at 14ème day and from 33.05, to 40.18 then to 42.35, and to 45.35 Pas at 21ème day and this for pectin incorporation rates of (0, 0.1, 0.3 and 0.6%) successively in yoghurts **(Table 04)**.

3.3. Microbiological quality of fermented milk supplemented with pectin 3.3.1. Streptococcus thermophilus

Overall, the evolution of the number of *Streptococcus thermophilus* in the experimental yoghurts was characterised by a clear increase during fermentation, with average values rising from 454 x 105 to 616 x 105 CFU/mL after 2 hours and 4 hours of fermentation, successively.

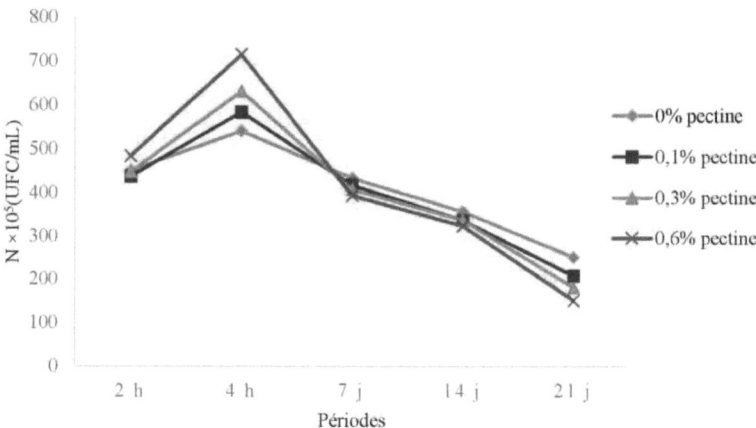

Figure 05. Variations in the number of *Streptococcus thermophilus* (CFU/mL) in yoghurts with pectin added.

During the post-acidification period, the average number of *Streptococcus thermophilus* in the experimental yoghurts fell from 412×10^5 CFU/mL on day $7^{\text{ème}}$ to 169×10^5 CFU/mL at the end of the storage period **(Figure 05)**.

At the end of the fermentation period, the number of germs appears to be higher as the rate of incorporation of pectin into the test products is increased ($p > 0.05$), with values varying from (540×105, to 583×105, to 630×105, and to 714×105 CFU/mL) for pectin rates of (0, 0.1, 0.3 and 0.6%) incorporated into yoghurts, respectively.

On the other hand, it was observed that during the storage period, the number of germs became inversely proportional to the pectin addition rates ($p > 0.05$); with values of (433×105, 417×105, 407×105, and 393×105 CFU/mL) at $7^{\text{ème}}$ day, (356×105, 336×105, 336×105, and 323×105 CFU/mL) at $14^{\text{ème}}$ day, and (250×10^5, 207×10^5, 180×10^5, and 150×10^5 CFU/mL) after 21 days of cold storage.

The analysis of variance shows a non-significant effect of the pectin incorporation rate on the average change in the number of *Streptococcus thermophilus* in the experimental fermented milks during the fermentation and post-acidification periods.

3.3.1. *Lactobacillus bulgaricus*

The number of *Lactobacillus bulgaricus* in yoghurts with pectin added increased from 397×10^5 CFU/mL at 2 hours to an average of 809×10^5 CFU/mL after 4 hours of steaming.

This increase continued until day $14^{\text{ème}}$ of storage, with an average value reaching (10×10^7 CFU/mL) at day $7^{\text{ème}}$, (10×10^7 CFU/mL) at day $14^{\text{ème}}$, followed by a decrease in the number of lactobacilli to (7×10^7 CFU/mL) at the end of the storage period at 4°C **(Figure 06)**.

During the 4 hours of fermentation, the number of *Lactobacillus bulgaricus* increased

18

according to the variable rates of 0, 0.1, 0.3 and 0.6% of pectin incorporated (p<0.05); with values that varied remarkably in the products and respectively from (710*10⁵ , to 800*10⁵ , to 824*10⁵ , and to 904*10⁵ CFU/mL) after 4 hours of fermentation.

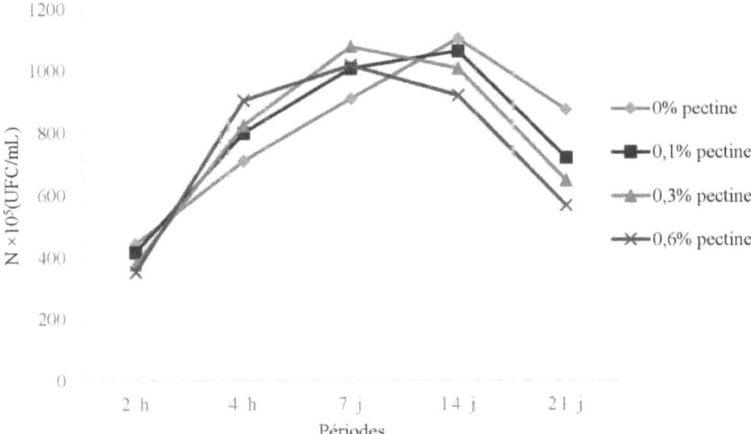

Figure 06. Variations in the number of *Lactobacillus bulgaricus* (10⁵ CFU/ml) in fermented milk with added pectin.

This trend continued until day 7ème of post-acidification, and even beyond, then an inversely proportional relationship was established between the number of germs and the level of added pectin, from the second week until the end of this conservation phase; with values fluctuating from (874*105, to 720*105, to 647*105, and to 567*105 CFU/mL) for variable pectin levels of (0, 0.1, 0.3 and 0.6%) in the experimental yoghurts, respectively.

The analysis of variance revealed a significant effect of the pectin incorporation rate on the average change in the number of *Lactobacillus bulgaricus* in the experimental yogurts during the fermentation and post-acidification periods.

3.4. Organoleptic quality of fermented milks supplemented with pectin 3.4.1. Taste :

During the post-acidification period, yoghurts with added pectin (0, 0.1, 0.3 and 0.6%) had rank sum values ranging from 31, 30, 19.5 and 19.5 rank sum at 1er day to 33.5, 31.5, 21.5 and 13.5 after 21 days of storage at 4°C, respectively. The results of (**Table 05**) show that yoghurts with 0.6% pectin added have the best taste values compared with yoghurts with 0, 0.1 and 0.3% pectin added.

During the post-acidification period, the jury rated the taste of the experimental yoghurts as good; the improvement in taste was proportional to the levels of pectin added (p<0.01). This was clearly observed in the second and third weeks of storage (p<0.01) (**Table 05**).

19

Table 05: Sensory evaluation of the taste (Rank Sum) of fermented milks with added pectin.

Period	Pectin incorporated (%)				Effect of pectin
	0 %	0.1 %	0.3 %	0.6 %	incorporation
1j	31 a	30 a	19.5 a	19.5 a	*
7j	28	29.5	22	20.51	NS
14j	30.5 a	29.5 a	24 ab	16 b	* *
21 j	33.5 a	31.5 a	21.5 b	13.5 c	* *

Results are expressed as rank sums; **: highly significant effect (p<0.01) of adding pectin; *: significant effect (p<0.05) of adding pectin; NS: non-significant effect (P>0.05) of adding pectin; a, b, c: statistical comparison of rank sums.

3.4.1. Cohesiveness :

During the storage phase, the evolution of the rank sums of the cohesiveness of the experimental yoghurts tended to increase with the increase in pectin levels (p<0.01), i.e. mean rank sums of (32.63, 35, 18 and 14.63) for pectin doses of (0, 0.1, 0.3 and 0.6%) incorporated into the products respectively **(Table 06)**.

Table 06. Sensory evaluation of the cohesiveness (rank sums) of fermented milks with added pectin.

Period	Pectin incorporated (%)				Effect of
	0%	0.1 %	0.3 %	0.6 %	incorporating pectin
1 j	30.5 a	32 a	20 b	17.5 b	* *
7 j	33.5 a	35.5 a	18 b	13 c	* *
14 j	33 a	36 a	17 b	14 b	* *
21 j	33.5 a	35.5 a	17 b	14 b	* *

Results are expressed as rank sums; **: highly significant effect (p<0.01) of adding pectin; a, b, c: statistical comparison of rank sums.

After 21 days of storage, yoghurts with 0.6% pectin added recorded the best cohesivity values (14 rank sums); on the other hand, those with 0, 0.1 and 0.3% added had poor cohesivity values of 33.5, 35.5 and 17 rank sums, respectively. The analysis of variance shows the highly significant effect of the pectin incorporation rate on the evolution of the cohesiveness of the experimental yogurts during cold storage of the products for 21 days of the post-acidification period.

3.4.2. Adhesiveness :

During the 21 days of cold storage, all the samples with varying amounts of pectin incorporated

20

(0.1, 0.3, and 0.6%) had average rank sum values of (26.25, 24, and 17) respectively compared with the control (0% pectin) for which an average value of (27.63) was recorded. Adhesiveness therefore seems to have improved with increasing pectin content in the products.

Statistical processing of the results reveals that the rate of pectin incorporation has a significant effect (p<0.05) on changes in the stickiness of experimental yoghurts during the post-acidification period (Table 07).

Table 07: Sensory evaluation of the stickiness (rank sums) of fermented milks with added pectin during the post-acidification period.

Periods	Pectin incorporated (%)				Effect of incorporating pectin
	0%	0,1 %	0,3 %	0,6 %	
1 j	17.5	20	23	19	NS
7 j	31[a]	26.5 [ab]	26.5 [ab]	16 [b]	*
14 j	31 [a]	28.5 [a]	24.5 [ab]	16 [b]	*
21 j	31 [a]	30 [a]	22 [ab]	17 [b]	*

Results are expressed as rank sums; *: Significant effect (p<0.05) of adding pectin; NS: Non-significant effect (p>0.05); a, b: Statistical comparison of rank sums.

3.4.3. Aftertaste :

Throughout the post-acidification period, the tasting panel concluded that the aftertaste was more pronounced the lower the levels of orange pectin incorporated into the products (p<0.01), with rank sum averages varying from (39.75, to 29.62, to 18.75, and to 11.87) for pectin levels varying from (0, 0.1, 0.3 and 0.6%) respectively in the products (Table 08).

Table 08: Sensory evaluation of the aftertaste (rank sums) of fermented milks with added pectin.

Period	Pectin incorporated (%)				Effect of pectin incorporation n
	0%	0,1 %	0,3 %	0,6 %	
1 j	40 [a]	29 [b]	20 [b]	11 [d]	* *
7 j	40 [a]	29.5 [b]	18.5 [c]	12 [d]	* *
14 j	39.5 [a]	30 [b]	18 [c]	12.5 [d]	* *
21 j	39.5 [a]	30 [b]	18.5 [c]	12 [d]	* *

Results are expressed as rank sums; **: highly significant effect (p<0.01) of adding pectin; NS: non-significant effect (p>0.05); a, b, c, d: statistical comparison of rank sums.

21

3.4.4. Whey exudation :

Throughout the cold storage period at 4°C, all the samples with pectin incorporated at (0.1, 0.3, and 0.6%) showed rank sum values varying on average from (29.38, to 24.88, and to 16.5) respectively. These values are still better compared to the standard control without pectin (0%) for which a rank sum level of 29.38 was recorded. Thus, the phenomenon of whey exudation proved to be inversely proportional ($p < 0.01$) to the increase in pectin doses in the experimental products **(Table 09)**.

The analysis of variance shows a highly significant effect ($p < 0.01$) of the pectin incorporation rate on changes in whey exudation from the experimental fermented milks during the post-acidification period.

Table 09: Sensory evaluation of whey exudation (rank sums) of fermented milks with added pectin.

Periods	Pectin incorporated (%)				Effect of incorporating pectin
	0%	0,1 %	0,3 %	0,6 %	
1 j	25.5	30	26	17.5	NS
7 j	23.5 [a]	33.5 [a]	28 [a]	15 [b]	* *
14 j	31 [a]	30 [a]	23.5 b[a]	17 [b]	*
21 j	37.5 [a]	24 [b]	22 [b]	16.5 [b]	* *

Results are expressed as rank sums; **: Highly significant effect ($p < 0.01$) of adding pectin; *: Significant effect ($p < 0.05$) of adding pectin; NS: Non-significant effect ($p > 0.05$); a, b: Statistical comparison of rank sums.

CHAPTER 4

4. Discussion

4.1. Chemical and techno-functional characteristics of pectins

The extraction yield of pectin is estimated at 24.33%. This yield is close to that reported by **Maran et al. (2013)** of around 19.24%. Whereas, **Zanella and Taranto (2015)** found very high extraction yields of orange albedo pectin from the *Citrus sinensis L. osbeck* variety approaching 38.21%. However, other work by **Guo et al (2012)** showed very low yields of around 15.47%. The extraction parameters (pH, time and temperature) as well as the characteristics of the raw material are at the origin of these variations **(Fishman et al., 2000)**. **Kalapathy and Proctor (2001)**, showed that a low temperature and a short extraction time lead to a low extraction yield, in addition the acid used for extraction and the nature of the alcohol used for precipitation can strongly influence the extraction yield of pectin. Heating the HCl solution allows hydrolysis of the pectic components located mainly in the middle layer of the cell walls (proto-pectin), which increases the pectin yield. **Chan and Choo (2013)** found that a low temperature is insufficient for the hydrolysis of protopectin (insoluble form of pectin) by acid, which results in a low yield of pectin.

Pectic polysaccharides are found mainly in the middle lamella between cells in the tissues of higher plants. They have a high molecular weight, and are tightly bound to the other polymers making up the cell walls, which prevents their release from the cell matrix. To extract the pectic substances contained in orange peel, microwave pre-treatment of the plant material has been recommended to facilitate pectin extraction **(Kratchanova et al., 2004; Rezzoug et al., 2008)**. **Kratchanova et al. (2004)** reported that during microwave pre-treatment, considerable pressure builds up inside the cells. This high pressure then alters the physical properties of orange peel tissue and its basic shape, breaking down the cell structure and enhancing the porous capillary structure of orange tissue. This device allows better penetration of the extraction solvent into the tissue, thereby improving pectin extraction, while considerably reducing extraction time.

The same authors also reported that the pre-treatment of fruit peels by microwave necessarily leads to a considerable increase in pectin yield and quality. This is due firstly to the partial disintegration of plant tissue and hydrolysis of protopectin, and secondly to the rapid inactivation of pectolytic enzymes **(Kratchanova et al., 2004)**.

In addition, **Yapo (2009)** reports that citric acid makes it possible to extract a less degraded pectin (without depolymerisation or de-esterification). As a result, it leads to pectin isolates with better gelling properties, which can be used in the agri-food sector even if they contain acid residues, without presenting any danger to consumers. On the other hand, extraction using strong acids, such as hydrochloric acid, results in the extraction of pectin that is more or less altered (depolymerisation

23

and/or de-esterification). Nevertheless, it remains the most widely used method on an industrial scale, due to the availability of mineral acids and their low cost. The use of strong acid solutions seems to be the most suitable, especially in view of the higher pectin extraction yields obtained compared with the use of citric acid **(Kanmani et al., 2014; Zanella and Taranto 2015)**.

Moisture levels appear to be low in our samples compared with commercial pectin. Moisture is a very important factor in pectin preservation. A low moisture content increases storage times and inhibits the growth of microorganisms that affect pectin quality through the production of hydrolytic enzymes (pectinases) **(Mohamadzadeh et al., 2010)**.

The ash content of experimental pectin is lower than that of commercial pectin. A low ash content is favourable for gel formation. The maximum limit of ash content for better quality pectin gels is 10% **(Ismail et al., 2012)**.

The equivalent weight of orange peel pectin used in the study is 620.03. **Kanmani et al. (2014)** found that the equivalent weight of pectin from *Citrus sinensis*, *Citrus limetta* and *Citrus limon* is 594.86, 386.45 and 253.70, respectively. These results clearly show the existence of a varietal difference in equivalent weight, with the highest value recorded for *Citrus sinensis*. The value of the equivalent weight of pectin can also vary according to the raw material and its degree of ripening. **Azad et al (2014)** found that the equivalent weight of pectin extracted from lemon varies by 1175, 1632 and 368 during three stages of ripening: before ripening, ripening and after ripening, respectively. According to the same authors, the ripening stage has an effect on significant ($p<0.05$) on the equivalent weight value. Samples extracted during the ripening phase had the highest equivalent weight, while the lowest value was recorded for samples from the post ripening phase. This decrease may be due to a partial degradation of the pectin.

The equivalent weight of pectin is also a function of the total content of free (non-esterified) galacturonic acid in the chain of the pectin molecule **(Rangama, 1977)**. According to **Rouse (1977)**, a higher degree of esterification causes a decrease in the free acid content and consequently leads to an increase in the equivalent weight value. The decrease in equivalent weight may be due to partial degradation of the pectin and depends on the amount of free acid **(Ramli and Asmawati, 2011)**.

The orange pectin extracted can be classified as a low methylation pectin (low methyl pectin) since its degree of esterification is less than 50%. These low-methyl pectins are often used in the food industry as gelling agents in low-sugar products such as low-calorie jellies and jams **(Tang et al., 2011)**.

Experimental orange peel pectin has a low methoxyl content and a lower degree of esterification than commercial pectin. Several studies have shown variable methoxyl contents in *Citrus sinensis*, *Citrus limetta* and *Citrus limon*; of 6.84%, 4.46% and 2.34%, respectively **(Kanmani et al., 2014)**. The same authors found in *Citrus sinensis*, *Citrus limetta* and *Citrus limon* species

24

esterification degrees of 3.20%, 2.98% and 1.50%, successively.

The methoxyl content of pectin can also vary depending on the plant species: mango peel 7.33%, banana (7.03%), grapefruit peel (3.57%) and lemon (9.92%) (**Madhav and Pushpalatha, 2002**). According to **Ismail et *al*. (2012)**, this content varies from 2.98% to 4.34%. However, **Azad et *al*. (2014)** showed that the methoxyl content of lemon pectin can vary from 4.26% to 10.25% depending on the ripening state of the fruit. The methoxyl content and degree of esterification also differ according to the extraction conditions (**Chan and Choo, 2013**). Methoxyl content is a very important factor in controlling the time and capacity of pectin gels to form (**Constella and Lozano, 2003**).

Weakly methylated pectin (DE<50%) can form a gel in the presence of bivalent ions, for example calcium ions, in the presence or absence of sugar (**Combo et *al*., 2011**). The degree of esterification (DE) and the distribution of free carboxyl groups are two important factors in the gelation of weakly methylated pectins. The lower the DE, the greater the affinity of the pectin chains for calcium ions (Ca^{2+}), resulting in stiffer gels (**Willats et *al*., 2006**). Highly methylated pectins (DE>50%) form a gel in the presence of sugar at concentrations greater than 55% (w/w) and in an acid medium (pH 2 - 3.5). In contrast, low-methyl pectins (DE<50%) require calcium ions (Ca^{2+}) to form a gel at pH ranging from 2.0 - 7.0, in the presence or absence of sugar (**Liu et *al*., 2010**).

The percentage of galacturonic acid (GA) is a very important factor in determining the purity of pectin. It is recommended that it should be greater than 65% (**Food Chemicals Codex, 1996**). However, the galacturonic acid (GA) content of the pectin we studied, extracted from orange peel, was less than 65%. These results indicate that this pectin is not pure. The same results were found by **Ismail et *al* (2012)**. The low galacturonic acid (GA) content may be due to the presence of sugar in the pectin precipitate. According to **Ismail et *al* (2012)**, a galacturonic acid (GA) percentage of less than 65% (low purity pectin) may be due to the presence of proteins, starch or sugar in the alcohol-precipitated pectin.

The nature of the acid used for extraction is an important factor that can also influence the galacturonic acid (GA) content of pectin. **Bhat and Singh (2014)** found that extraction of pectin using hydrochloric acid leads to a lower GA content than pectin extracted using citric acid. According to the same authors, this low content may be due to the presence of sugars in the pectin precipitate. Similarly, **Mohamed and Mohamed (2015)** showed that the GA content is higher when pectin is extracted using a hydrochloric acid solution (33.90%), followed by extraction using water (31.8%) and finally, pectin extracted using ammonium oxalate (27.7%). The GA content also varies according to the raw material used for pectin extraction (extraction source). **Girma and Worku (2016)** found that the galacturonic acid (GA) content of pectin extracted from mango (70.65%) was higher than that of banana pectin (53.60%).

25

4.2. Quality of yoghurts with added pectin

During the fermentation and post-acidification periods, an increase in acidity proportional to pectin levels was recorded. According to **Luquet (1990),** such results can only be justified by the production of lactic acid due to fermentation of the lactose constituting the milk by the specific microorganisms inoculated. The higher the rate of incorporation of pectin into the medium, the higher the lactate content. This suggests that the pectin acts by stimulating the fermentation activity of the specific yoghurt germs, resulting in intense production of lactate in the medium. Furthermore, the main products of lactic bacteria metabolism are made up of several organic acids, which are produced either by the homofermentative pathway (lactic acid only), or by the heterofermentative pathway (lactic, acetic and formic acids) and which can also acidify and vary the pH of the medium **(Combo et al., 2011).**

The increase in acidity of the experimental products began as early as the first week of cold storage, and seemed to stabilise until day 21ème . This is probably due to the phenomenon of gelation caused by the addition of pectin to the yoghurts, which has the ability to complex the free water in the medium, lowering the Aw necessary for the development of the bacteria inoculated and therefore their ability to ferment the lactose into lactic acid **(Buléon et al., 1998).**

These results also reflect the consistency of the pH values obtained, which are inversely proportional to Dornic acidity on the one hand, and to the levels of pectin incorporated into the fermented milks on the other.

The reduction in pH and increase in lactic acidity is due to the fermentation of milk lactose by the two specific strains *Streptococcus thermophilus* and *Lactobacillus bulgaricus*. **Sokolinska et al (2004)** found that the pH of fermented milk fell from 6.7 to 4.11 during the fermentation and post-acidification period.

Streptococcus thermophilus and *Lactobacillus bulgaricus* live in symbiosis and there is a synergy between the two bacteria that involves mutual stimulation. This stimulation mainly involves growth, acidification and the production of aromatic compounds. *Streptococcus thermophilus* is stimulated by the supply of amino acids and small peptides from the proteolytic activity of *Lactobacillus bulgaricus*. The stimulation of *Lactobacillus bulgaricus* is attributed to formic acid, pyruvic acid and carbon dioxide produced by *Streptococcus thermophilus*. Both microbial species are homo-fermentative bacteria that produce lactic acid from milk lactose. The production of lactic acid leads to a lowering of the pH. As they approach isoelectric pH (pHi 4.6), casein micelles lose their steric stability, causing them to flocculate, precipitate and form a coagulum **(Loveday et al., 2013).** **Kumar and Mishra (2004),** also found that the lactic acidity of yoghurt increased with increasing pectin addition rate from 0.2, to 0.4 and 0.6%.

In terms of texture, the experimental yogurts were characterised by a clear increase in viscosity during both the fermentation and post-acidification phases. This viscosity can be described as the resistance shown by a standardised ball as it moves through a liquid (**Schroder et al., 2004**). According to **Rawson and Marshall, 1997**, this is related to the ability of the seeded strains to produce exopolysaccharides (EPS), particularly *Streptococcus thermophilus* during the fermentation phase when they are most active. These exopolysaccharides increase the viscosity and improve the texture of fermented milks (**Cerning, 1995**). According to **Girard and Lequart (2007)**, specific yoghurt germs, particularly *Streptococcus thermophilus,* produce excpolysaccharides during lactic fermentation that are able to bind to milk casein, giving the finished product a particular viscosity and rheological quality. **Guzel-Seydim et al (2005)** found that the viscosity of fermented milks prepared by exopolysaccharide-producing bacteria is often much higher than those prepared by bacteria unable to produce them.

Increasing the pectin incorporation rate in fermented milks from 0, to 0.1, to 0.3 and to 0.6% is accompanied by a significant increase ($p < 0.01$) in viscosity. The same results were obtained by **Jensen et al (2010)** who found that increasing the pectin concentration from 0.2 to 0.5% resulted in an increase in the viscosity of acidified milks. EPS production and viscosity therefore appear to be closely proportional to the doses of pectin incorporated. These results can be explained by the fact that the incorporated pectin can form a three-dimensional network capable of complexing the milk constituents while absorbing as much water as possible from the medium, resulting in an increase in the viscosity of the experimental yoghurts (**Moll and Moll, 1998**).

Pectin is essentially made up of galacturonic acid residues linked together by a-bonds (1^4), partially acetylated or esterified by methyl groups, which gives it its gelling, thickening and stabilising properties, and it also has a very high water retention capacity (**Fishman et al., 2000**). In this way, pectin can form a three-dimensional network capable of complexing milk constituents while absorbing a maximum amount of water from the medium, resulting in an increase in the viscosity of experimental yoghurts (**Maroziene and Kruif, 2000**). Once absorbed onto the surface of casein micelles, pectin can form stable aggregates (**Maroziene et Kruif, 2000 ; Tuinier et al., 2002 ; Kiani et al., 2010**).

In the same context, the organoleptic parameters relating to yoghurt texture, stickiness and cohesiveness, can be understood in a similar way to viscosity; bearing in mind that the gel formed is a mixture of pectin and casein, and therefore its strength and these rheological criteria increase proportionally with the levels of pectin incorporated (**Laurent and Boulenguer, 2003**).

According to **Bourgois et al (1989)**, *Streptococcus thermophilus* species are responsible for starting lactic fermentation, and grow up to a certain pH of the medium (4.2); above this value, these germs are inhibited and *Lactobacillus bulgaricus* take over, completing the fermentation. Similarly, **Guyot (1992)** reports that *Streptococcus thermophilus* start the lactic fermentation of yoghurts, and

27

their growth is stimulated by the amino acids released as a result of the proteolytic activity of *Lactobacillus bulgaricus* from the milk caseins. This results in a higher number of *Streptococcus thermophilus* during the first incubation phase, followed by an inhibitory effect of lactic acid on *Streptococcus thermophilus*, leading to a drop in their numbers **(Jeantet et al., 2008)**.

Generally speaking, the number of these two strains appeared to be proportional to the pectin concentrations. A stimulatory effect of pectin was noted, particularly with regard to the growth of *Lactobacillus bulgaricus* during the post-acidification phase. Thanks to the use of pectins as a carbon source **(Olano-Martin et al., 2002; Manderson et al., 2005; Combo et al., 2011)**, *Lactobacillus bulgaricus* showed a marked increase in proportion to the increase in pectin concentrations in the experimental products. **Kumar and Mishra (2004)** eventually found that the rate of incorporation of pectin in yoghurt had a significant effect on the growth of the two strains: *Streptococcus thermophilus* and *Lactobacillus bulgaricus*.

These results are confirmed by **Buléon et al (1998)**, who report that the action of the gel formed following the addition of pectin during preservation can affect the water activity (aW) of the products, which can result in partial inhibition of the growth of the germs sown.

Organoleptic quality was significantly improved by increasing the rate of pectin incorporation in the experimental yogurts. In fact, at the higher pectin levels, the products showed a better taste, a firmer gel and texture, even limiting whey exudation, while the aftertaste of the fermented milks seemed to be more pronounced. As regards stickiness and cohesiveness, the panelists noted a clear improvement in these criteria as a function of the levels of pectin added. Similar results were obtained by **Kumar and Mishra (2004)** who found an improvement in the stickiness and cohesiveness of experimental fermented milks with the addition of pectin.

The gel formed is a mixture of pectin and casein, the strength of which is proportional to the level of additive incorporated **(Laurent and Boulenguer, 2003)**. This is confirmed by **Jensen et al (2010)** who report that increasing the concentration of pectin from 0.2 to 0.5% results in a remarkable increase in the elastic and viscous properties of pectin gels. In the same context, the results of **Broomes and Badrie (2010)** show the significant effect of pectin in producing a firmer gel texture. In addition, it is well established that specific yoghurt germs, particularly *Streptococcus thermophilus*, produce exopolysaccharides (EPS) in the environment during the first phase of fermented milk production. These are carbohydrate substances, particularly P-glucan, which can bind to milk caseins during fermentation while improving the rheological quality of yoghurts **(Lorient et al., 1985; Cerning et al., 1986; Rawson and Marshall, 1997)**.

According to **Bottazzi et al (1973)**, consumer appreciation of the flavour and taste of fermented milks can be just as important as consistency and smoothness. These parameters are clearly improved in proportion to the doses of pectin incorporated into the products. Apparently, pectin can stimulate specific yoghurt bacteria to produce more acetaldehyde **(Soukoulis et al., 2007)**, which is responsible for the characteristic taste of yoghurt. The acetaldehyde formed during lactic fermentation is the main component of the specific flavour of yoghurt **(Sahan et al., 2008)**. The interaction between milk proteins and pectin leads to unfolding of the proteins, making hydrophobic groups accessible. These groups provide additional binding sites for volatile compounds **(Mao et al., 2014)**. This leads to a reduction in the volatility of flavour compounds and therefore a better flavour of fermented milks with added pectin. In addition, the increase in viscosity can influence the mobility of flavour compounds within the matrix **(Mao et al., 2014), which** can reduce their release into the gas phase and the resulting olfactory perception is improved.

The improvement in texture, due to the incorporation of pectin, has resulted in an apparent limitation of the phenomenon of product syneresis defined as the separation of whey from curd without the application of an external force **(Peng et al., 2009) during** the storage period **(Zare et al., 2011)**. Indeed, pectins are anionic hydrocolloids capable of interacting with the positive charges on the surface of proteins (caseins, serum proteins), reinforcing the three-dimensional network and thus controlling syneresis **(Soukoulis et al., 2007)**. Whey exudation was inversely proportional to increasing doses of pectin incorporated into experimental yoghurts ($p < 0.01$). Similar results were reported by **Everett et al (2005)**, who suggested that the addition of pectin to yoghurt improves whey exudation due to the absorption of pectin on the surface of milk casein micelles, which consequently increases the water retention capacity of fermented milks.

CHAPTER 5

Conclusion

Today, the by-products of the agri-food industry are a major source of pollution and high economic losses. Getting the most out of these by-products has become a key requirement for both economic and environmental reasons.

Pectin is a complex polysaccharide that forms part of the cell walls of most higher plants. Pectins are abundant in fruit and vegetables. Although they can be extracted from a large number of plants, the main industrial sources of pectins are apple pomace and orange peel. There are many applications for this substance in various fields, but the most important use is in the food industry, where pectins are mainly used as texturising agents, stabilisers, gelling agents and thickeners.

In the light of the results obtained, it appears that during the fermentation and post-acidification period, the Dornic acidity values recorded are proportional to the levels of pectin added (0, 0.1, 0.3, and 0.6%) to the yoghurts; whereas an inverse evolution of the pH values as a function of the levels of pectin incorporated into the products was observed.

Furthermore, during the post-acidification storage period, the average acidity content of yoghurts with pectin added seemed to increase slightly until the last day of cold storage at $21^{\text{ème}}$, but without exceeding the standards allowed by the regulations.

It should also be noted that viscosity showed an interesting improvement during fermentation, especially in the product prepared at a concentration of 0.6% pectin. Furthermore, during the post-acidification period, this trend seems to be maintained; the viscosity of the fermented milks showed a clear change in proportion to the levels of pectin incorporated.

A greater proliferation of *Streptococcus thermophilus* germs than *Lactobacillus bulgaricus* was observed during fermentation, in contrast to the post-acidification period. The higher the pectin content in the medium, the greater the proliferation of these germs during both periods. The number of *Streptococcus thermophilus* and *Lactobacillus bulgaricus* in the fermented milk complied with the accepted standard for yoghurt of 10^8 live bacteria/mL of product.

Organoleptic quality showed an improvement in acid taste and aftertaste with the increase in pectin content in yoghurts. Pectin incorporated at levels of 0.1, 0.3 and 0.6% clearly improved the rheological quality, particularly the viscosity, adhesiveness and cohesiveness of fermented milks. Yoghurts with 0.6% pectin recorded the best values for cohesiveness, stickiness and taste. Pectin significantly improved the keeping qualities, even limiting whey exudation.

References

-A-

1. **AFNOR, 1980**. Lait et produits laitiers : méthodes d'analyse (1er éd.). Paris: AFNOR.

2. **AOAC, 1980.** Association of official analytical chemists -Official Methods of Analysis. 13th ed., Washington D.C.

3. **AOAC, 2005.** Association of official analytical chemists- official methods analysis of the association analytical chemists (18th ed.). Washington, DC: AOAC.

4. **Atmodjo M. A., Hao Z., Mohnen D., 2013.** Evolving Views of Pectin Biosynthesis. *Annual Review of Plant Biology*, 64:747-779.

5. **Azad A. K. M., Ali M. A., Akter M. S., Rahman M. J., Ahmed M., 2014.** Isolation and characterization of pectin extracted from lemon pomace during ripening. *Journal of Food and Nutrition Sciences*, 2(2): 30-35.

-B-

6. **Bhat S. A., Singh E. R., 2014.** Extraction and characterization of pectin from guava fruit peel. *International Journal of Research in Engineering and Advanced Technology*, 2 (3):1 - 7.

7. **Bhatia M.S., Deshmukh R., Choudhari P., Bhatia N.M., 2008.** Chemical Modification of Pectins, Characterization and Evaluation for Drug Delivery. *Scientia Pharmaceutica*, 76: 775-784.

8. **Bottazzi V., Battistotti B. and Montescani G., 1973.** Influence of single and combined strains of *L. bulgaricus* and *Str. thermophilus* and of milk treatments on the production of acetic aldehyde in yoghurt. *Mémoires Originaux, Le Lait*, No. 525-526 (May-June): 295 - 308.

9. **Bourgois C.M., Larpent J.P., 1989.** Food microbiology. *Ed. Lavoisier, Technique & Documentation, Vol.* 2. pp : 18 - 30.

10. **Braccini I., Perez S., 2001.** Molecular basis of Ca^{2+} -induced gelation in alginates and pectins: The egg-box model revisited. *Biomacromolecules*, 2: 1089-1096.

11. **Broomes J., Badrie N., 2010.** Effects of Low-Methoxyl Pectin on Physicochemical and Sensory Properties of Reduced- Calorie Sorrel/ Roselle (*Hibiscus sabdariffa* L.) Jams. *The Open Food Science Journal*, 4: 48-55.

12. **Buléon A., Coloma P., Bail P., Bobol H., 1998.** Structure and phase transition of starches: Application à la formation. *Ed. INRA.* pp : 2 - 15.

-C-

13. **Caffall K. H., Mohnen D., 2009.** The structure, function and biosybthesis of plant cell wall pectic polysaccharides. *Carbohydrate Research*, 344:1879-1900.

14. **Cerning J., 1995.** Production of exopolysaccharides by lactic acid bacteria and dairy propioni bacteria. *Lait*, 75: 463 - 472.

15. **Cerning J., Bouillance C., Desmazeaud M.J., 1985.** Isolation and characterisation of exocellular polysaccharides produced by *Lactobacillus bulgaricus*. *Biotechnology Letter*

Scientific, 8: 6 - 25.

16. **Chan S.Y., Choo W.S., 2013.** Effect of extraction conditions on the yield and chemical properties of pectin from cocoa husks. *Food chemistry*, 141: 3752-3758.

17. **Combo A. M. M., Aguedo M., Paquot M., 2011.** Pectic oligosaccharides: production and possible applications. *Biotechnol. Agron. Soc.*, **15**(1): 153-164.

18. **Constenla D., Lozano J. E., 2003.** Kinetic model of pectin demethylation. *Latin American Applied Research*, 33:91-96.

-F-

19. **Fishman M.L., Chau H.K., Hoagland P., Ayyad K., 2000.** Characterization of pectin, flash-extracted from orange a+lbedo by microwave heating, under pressure. *Carbohydrate Research*, 323: 126- 138.

20. **Food Chemical Codex. 1996.** IV monographs.Washington DC: National Academy Press, pp : 283.

-G-

21. **Girard M., Lequart C.S., 2007.** Gelation and resistance to schearing of fermented milk: Role of exopolysaccharides. *International Dairy Journal*, 17 : 666- 673.

22. **Girma E., Worku T., 2016.** Extraction and characterization of Pectin from Selected Fruit Peel Waste. *International Journal of Scientific and Research Publications*, 6 (2):447- 454.

23. **Guo X., Han D, Xi H., Rao L., Liao X., Hu X., Wu J., 2012.** Extraction of pectin from navel orange peel assisted by ultra-high pressure, microwave or traditional heating: A comparison. *Carbohydrate polymers*, 88 : 441-448.

24. **Guyot P., 1992.** Les yaourts D.L.G., *Food Technology,* p: 4 - 8 - 10 - 11.

25. **Guzel-Seydim Z.B., Sezgin E., Seydim A.C., 2005.** Influences of exopolysaccharides producing cultures on the quality of plain set type yogurt. *Food control*, 16: 205- 209.

-I-

26. **Iglesias M.T., Lozano J.E., 2004.** Extraction and characterization of sun flower pectin. *Journal of Food Engineering*, 62: 215-223.

27. **International Dairy Federation, 2003.** Yoghurt: Enumeration of characteristic microorganisms -colony count technique at 37°C. IDF Standard No 117 E. Brussels.

28. **Ismail N. S. M., Ramli N, Hani N. M., Meon Z., 2012.** Extraction and Characterization of Pectin from Dragon Fruit (*Hylocereus polyrhizus*) using Various Extraction Conditions. *Sains Malaysiana,* 41(1): 41-45.

-J-

29. **Jeantet R., Croguennec T., Mahaut M., Schuck P., Brule G., 2008.** Les produits laitiers. *Technique et Documentation, Lavoisier*, Paris, p: 1-36.

30. **Jensen J.K.,S0rensen S. O., Harholt J., Geshi N., Sakuragi Y., M0ller I., Zandleven J., Adriana J. Bernal A. J., Jensen N. B., S0rensen C., Pauly M., Beldman G., Willats W.G.T., Henrik Vibe Schellera H. V.,2008.** Identification of a Xylogalacturonan Xylosyltransferase Involved in Pectin Biosynthesis in Arabidopsis. *The Plant Cell*, 20: 1289-1302.

31. **Jensen S., Rolin C., Ipsen R., 2010.** Stabilization of acidified skimmed milk with HM-pectin. Food Hydrocolloids, 24: 291- 299.

-K-

32. **Kalapathy U., Proctor A., 2001.** Effect of acid extraction and alcohol precipitation conditions on the yield and purity of soy hull pectin. *Food chemistry*, 73: 393- 396.

33. **Kanmani P., Dhivya E., Aravind J., Kumaresan K., 2014.** Extraction and Analysis of Pectin from Citrus Peels: Augmenting the Yield from *Citrus limon* Using Statistical Experimental Design. *Iranica Journal of Energy and Environment*, 5 (3): 303-312.

34. **Kar F., Arslan N., 1999.** Effect of temperature and concentration on viscosity of orange peel pectin solutions and intrinsic viscosity-molecular weight relationship. *Carbohydrate Polymers*, 40: 277-284.

35. **Keogh, M.K., O'Kennedy, B.T., 1998.** Rheology of stirred yogurt as affected by added milk fat, protein, and hydrocolloids, *Journal of Food Science*, 63 (1): 108-112.

36. **Kiani H., Mousavi M.E., Razavi H., Morris E.R., 2010.** Effect of gellan, alone and in combination with high-methoxy pectin, on the structure and stability of doogh, a yogurt-based Iranian drink. *Food hydrocolloids*, 24: 744- 754.

37. **Kratchanova M., Pavlova E., Panchev I., 2004.** The effect of microwave heating of fresh orange peels on the fruit tissue and quality of extracted pectin. Carbohydrate Polymers, 56, 181-185.

38. **Kravatchenko T., Voragen A., Pilnik W., 1992.** Analytical Comparison Of Three Industrial Pectin Preparations. *Carbohydrate Polymers*, 18: 17-25.

39. **Kumar P., Mishra H.N., 2004.** Mango soy fortified set yoghurt: effect of stabilizer addition on physicochemical, sensory and textural properties. *Food chemistry*, 87: 501- 507.

-L-

40. **Laurant M.A., Boulanguer P., 2003.** Stabilization mechanism of acid dairy drinks (ADD) induced by pectin. *Food Hydrocolloids*, 17: 445 - 454.

41. **Lira-Ortiz A. L., Reséndiz-Vega F., Ri'os-Leal E., Contreras-Esquivel J. C. , Chavarria-Hernandez N., Vargas-Torres A. Rodriguez-Hernandez A. I., 2014.** Pectins from waste of prickly pear fruits (*Opuntia albicarpa* Scheinvar 'Reyna'): Chemical and rheological properties. *Food hydrocolloids*, 37: 93-99.

33

42. **Liu L., Cao J., Huang J., Cai Y., Yao J., 2010.** Extraction of pectins with different degrees of esterification from mulberry branch bark. *Bioresource Technology,* 101 : 3268-3273.

43. **Lorient D., Cheftel J.C., Luquet J.L., 1985.** Food Properties, Biochemistry, Functional Properties, Nutritional Values, Chemical Modifications. *Ed. Lavoisier, Technique & Documentation,* pp: 39 - 53.

44. **Loveday S.M., Sarkar A., Sing H., 2013.** Innovative yoghurts: Novel processing technologies for improving acid milk gel texture. *Food Science and Technology*, 33: 5- 20.

45. **Lucey J. A., Singh H., 1998.** Formation and physical properties of acid milk gels: a review. *Food Research International*, 30 (7): 529- 542.

46. **Luquet F.M.,1990.** Lait et produits laitiers : Vache - Brebis - Chèvre. *Techniques et Documentation, Lavoisier*, Paris.

-M-

47. **MacDonald I., 1979.** Polysaccharides and Health. *In*: Polysaccharides in Food, Blanchard J.M.V., Mitchell J.R. (*Eds*), *Butterworths*, London, UK, Chapter 21, pp. 331-336.

48. **Madhav A, Pushpalatha P. B., 2002.** Characterization of pectin extracted from different fruit wastes. *Journal of Tropical Agriculture*, 40: 53-55.

49. **Manderson K. et al., 2005.** *In vitro* determination of the prebiotic properties of oligosaccharides derived from an orange juice manufacture by-product stream. *Appl. Environ. Microbiol*, 71(12): 8383-8389.

50. **Mao L., Bioteux L., Roos Y.H., Miao S., 2014.** Evaluation of volatile characteristics in whey protein isolate-pectin mixed layer emulsion under different environmental conditions. *Food Hydrocolloids*, 41: 79-85.

51. **Maran J.P., Sivakumar V., Thirugnanasambandham K, Sridhar R., 2013.** Optimization of microwave assisted extraction of pectin from orange peel. *Carbohydrate polymers,* 97 (2): 703-709.

52. **Maroziene A., Kruif C.G., 2000.** Interaction of pectin and casein micelles. *Food hydrocolloids*, 14: 391- 394.

53. **Mesbahi G., Jamalian J., Farahnaky A., 2005.** A Comparative Study on Functional Properties of Beet and Citrus Pectins in Food Systems. *Food Hydrocolloids, Vol.* 19 (4): 731-738.

54. **Mohamadzaheh J., Sadeghi-mahoonak A.R., Yaghbani M., Aalami M., 2010.** Extraction of pectin from sunflower head residues of selected Iranian cultivars. *World Applied Sciences Journal*, 8 (1): 21-24.

55. **Mohamed H. A., Mohamed B. E., 2015.** Fractionation and Physicochemical Properties of Pectic Substances Extracted from Grapefruit Peels. *Journal of Food Process Technology,* 6:

1 - 6.

56. **Mohnen D., 2008.** Pectin structure and biosynthesis. *Current Opinion in Plant Biology*, 11:266 - 277.

57. **Moll M. and Moll N., 1998.** Food additives and processing aids. *Dunod*, Paris.

-O-

58. **Olano-Martin E., Gibson G.R., Rastall R.A., 2002.** Comparison of the in vitro bifidogenic properties of pectins and pectic-oligosaccharides. *Journal of Applied Microbiology, 93:* 505-511.

59. **Owens H.S., McCready R.M., Shepard A.D., Schultz T.H., Pippen E.L., Swenson H.A., Miers J.C., Erlandsen R.F. Maclay, W.D., 1952.** Methods used at Western Regional Research Laboratory for Extraction of Pectic Materials. pp. 9. USDA Bur. Agric. Ind. Chem.

-P-

60. **Pang Z., Deeth H., Prakash S., Bansal N., 2016.** Development of rheological and sensory properties of combinations of milk proyeins and gelling polysaccharides as potential gelation replacements in the manufacture of stirred acid milk gels and yogurt. *Journal of Food Engineering*, 169: 27-37.

61. **Peng Y., Serra M., Horn D.S., Lucey J.A., 2009.** Effect of fortification with various types of milk protein on the rheological properties and permeability of nonfat set yogurt. *Journal of food Science*, 74 (9): C666 - C573.

62. **Pinheiro E. R., Silva I.M.D.A., Gonzaga L. V., Amante E. R., Teo'filo R. F., Ferreira M. M. C., Amboni R. D.M.C., 2008.** Optimization of extraction of high-ester pectin from passion fruit peel (*Passiflora edulis flavicarpa*) with citric acid by using response surface methodology. *Bioresource Technology*, 99 : 5561-5566.

-R-

63. **Ramli N , Asmawati, 2011.** Effect of ammonium oxalate and acetic acid at several extraction time and pH on some physicochemical properties of pectin from cocoa husks (*Theobroma cacao*). *African Journal of Food Science, 5*(15): 790-798.

64. **Ranganna S., 1977.** Manual of analysis of fruit and vegetable products. McGraw Hill, New Delhi.

65. **Rawson H.L., Marshall M., 1997.** Controlling the rheological properties of yoghurts. *Lettres Scientifiques et Techniques de TEXEL* No. 1. P : 4.

66. **Rezzoug S.A., Maache-Rezzoug Z., Sannier F., Allaf K., 2008.** A thermo mechanical preprocessing for pectin extraction from peel. Optimization by response surface methodology. *International Journal of Food Engineering*, Vol 4 (1), Article 10.

67. **Ridley B. L., O'Neill M. A., Mohnen D., 2001.** Pectins: structure, biosynthesis, and

oligogalacturonide-related Signaling. *Phytochemistry* ,57: 929-967.

68. **Rouse A.H., 1977.** Pectin: distribution, significance. Dalam Nagy SP, Shaw E, Veldhuis MK (eds). Citrus Science and Technol (1). The AVI Publishing Company Inc.

-S-

69. **Sahan N., Yasar K., Hayaloglu A.A., 2008.** Physical, chemical and flavour quality of non-fat yogurt as affected by a B-glucan hydro colloidal composite during storage. *Food hydrocolloids*, 22 : 1291- 1297.

70. **Sahan N., Yasar K., Hayaloglu A.A., 2008.** Physical, chemical and flavour quality of non-fat yogurt as affected by a B-glucan hydro colloidal composite during storage. *Food hydrocolloids*, 22 : 1291- 1297.

71. **Schroder R., Clark C.J., Sharrock K., Hallett I.C., MacRae E.A., 2004.** Pectins from the albedo of immature lemon fruitlets have high water binding capacity. *Journal Plant Physiology*, 161: 371-379.

72. **Shaha R.K., Punichelvana Y. N. A.P., Afandi A., 2013.** Optimized Extraction Condition and Characterization of Pectin from Kaffir Lime (*Citrus hystrix*). *Research Journal of Agriculture and Forestry Sciences, Vol.* 1(2): 1-11.

73. **Sodini I., Remeuf F., Haddad S., Corrieu G., 2004.** The relative effect of milk base starter and process on yoghurt texture: a review. *Food Science and Nutrition*, 44:113- 137.

74. **Sokolinska D.C., Mchalski M.M., Pikul J., 2004.** Role of the proportion of yoghurt bacterial strains in milk souring and the formation of curd qualitative characteristics. *Bull. Vet. Inst. Pulawy*, 48 : 437- 441.

75. **Soukoulis C., Panagiotidis P., Koureli R., Tzia C., 2007.** Industrial Yogurt Manufacture: Monitoring of Fermentation Process and Improvement of Final Product Quality. *Journal of Dairy Science, Vol. 90 (6):* 2641-2654.

76. **Srivastava P., Malviya R., 2011.** Source of pectin, extraction and its application in pharmaceutical industry: a review. *Indian Journal of Natural Products and Resources*, 2 (1): 10-18.

-T-

77. **Tang P. Y., Wong C. J., Woo K. K., 2011.** Optimization of pectin extraction from peel of dragon fruit (*Hylocereus polyrhizus*). *Asian Journal of Biological Sciences*, 4 (2) : 189 - 195.

78. **Tuinier R., Rolin C., Kruif C.G., 2002.** Electrosorption of pectin onto casein micelles. *Bio macromolecules*, 3: 632- 638.

-W-

79. **Wang S., Chen F., Wu J., Wang Z., Liao X., Hu X., 2007.** Optimization of pectin extraction assisted by microwave from apple pomace using response surface methodology. *Journal of*

Food Engineering, 78: 693-700.

80. **Wicker L., Kim Y., Kim M. J., Thirkield B., Lin Z., Jung J., 2014.** Pectin as a bioactive polysaccharide extracting tailored function from less. *Food Hydrocolloids*, 1-9.

81. **Willats W.G. T., Knox J. P., Mikkelsen J. D., 2006.** Pectin: new insights into an old polymer are starting to gel. *Trends in Food Science & Technology*, 17: 97-104.

82. **Willats W.G.T., McCartney L., Mackie W., Knox J.P., 2001.** Pectin: Cell Biology and Prospects For Functional Analysis. *Plant Molecular Biology*, 47: 9-27.

-Y-

83. **Yang T., Wu K., Wang F., Liang X., Liu Q., Li G., Li Q., 2014.** Effect of exopolysaccharides from lactic acid bacteria on the texture and microstructure of buffalo yoghurt. *International Dairy Journal*, 34: 252-256.

84. **Yapo B. M., 2009.** Pectin Quantity, Composition, and Physicochemiocal Behavior as Influenced by the Purification Process. *Food Reasech International*, 42: 1197 - 1202.

-Z-

85. **Zanelle K., Taranto O.P., 2015.** Influence of the drying operating conditions on the chemical characteristics of citric acid extracted pectins from pera sweet orange (Citrus Sinensis L. Osbeck) albedo and flavedo. *Journal of Food Enginneering*, 166 : 111-118.

86. **Zare F., Boye J.I., Orsat V., Champagne C., Simpson B.K., 2011.** Microbial, physical and sensory properties of yogurt supplemented with lentil flour. *Food Research International*, 44: 2482-2488.

Printed by Books on Demand GmbH, Norderstedt / Germany